Planets And Moons In Our Universe

Speedy Publishing LLC
40 E. Main St. #1156
Newark, DE 19711

www.speedypublishing.com

Copyright 2014
9781680321210
First Printed November 17, 2014

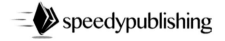

Did you know?

The Universe is all of spacetime and everything that exists therein, including all planets, stars, galaxies, the contents of intergalactic space, the smallest subatomic particles, and all matter and energy.

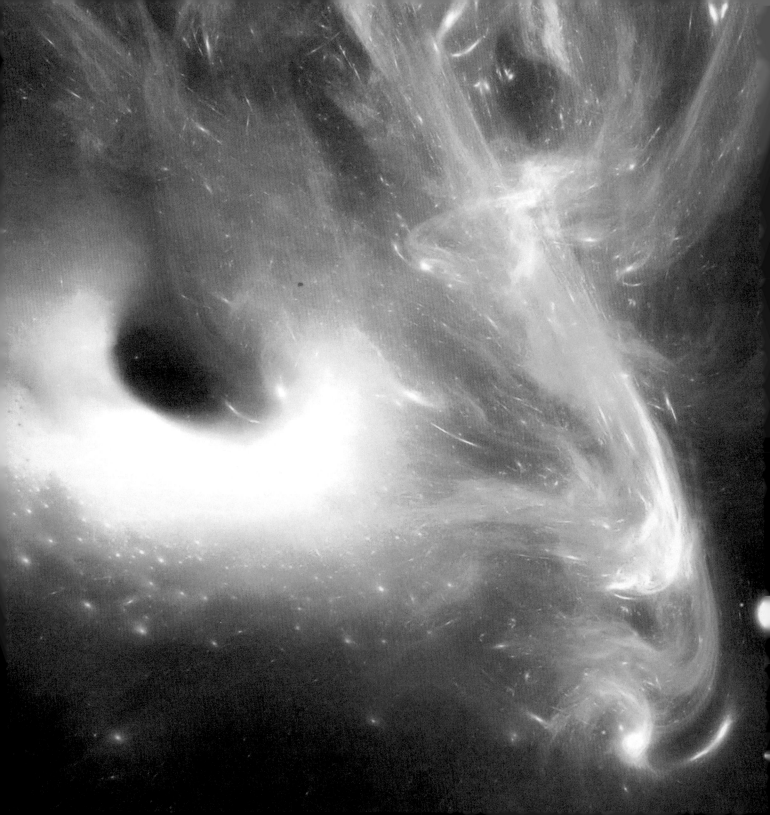

Did you know?

A galaxy is a massive, gravitationally bound system consisting of stars, stellar remnants, an interstellar medium of gas and dust, and dark matter, an important but poorly understood component. The Milky Way is one of billions of galaxies in the universe. There are as many galaxies as there are stars in the Milky Way.

Did you know?

Galaxies contain varying numbers of planets, star systems, star clusters and types of interstellar clouds. In between these objects is a sparse interstellar medium of gas, dust, and cosmic rays. Many galaxies are believed to have supermassive black holes (SMBH) at their center.

Did you know?

A planet, in astronomy, is one of a class of celestial bodies that orbit stars. The term planet is ancient, with ties to history, science, mythology, and religion. The planets were originally seen by many early cultures as divine, or as emissaries of deities.

Did you know?

The Solar System comprises the Sun and the objects that orbit it, whether they orbit it directly or by orbiting other objects that orbit it directly. Of those objects that orbit the Sun directly, the largest eight are the planets that form the planetary system around it, while the remainder are significantly smaller objects, such as dwarf planets and small Solar System bodies (SSSBs) such as comets and asteroids.

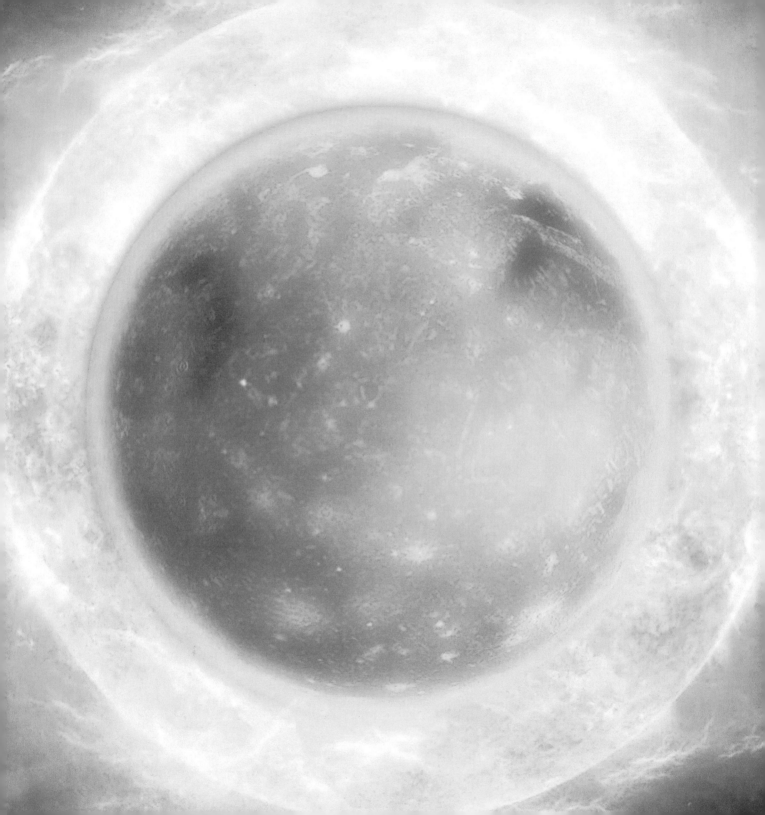

Did you know?

The Sun is the star at the center of the Solar System. It is almost spherical and consists of hot plasma interwoven with magnetic fields. It has a diameter of about 1,392,684 km (865,374 mi), around 109 times that of Earth, and its mass (1.989 × 1030 kilograms, approximately 330,000 times the mass of Earth) accounts for about 99.86% of the total mass of the Solar System. Chemically, about three quarters of the Sun's mass consists of hydrogen, whereas the rest is mostly helium.

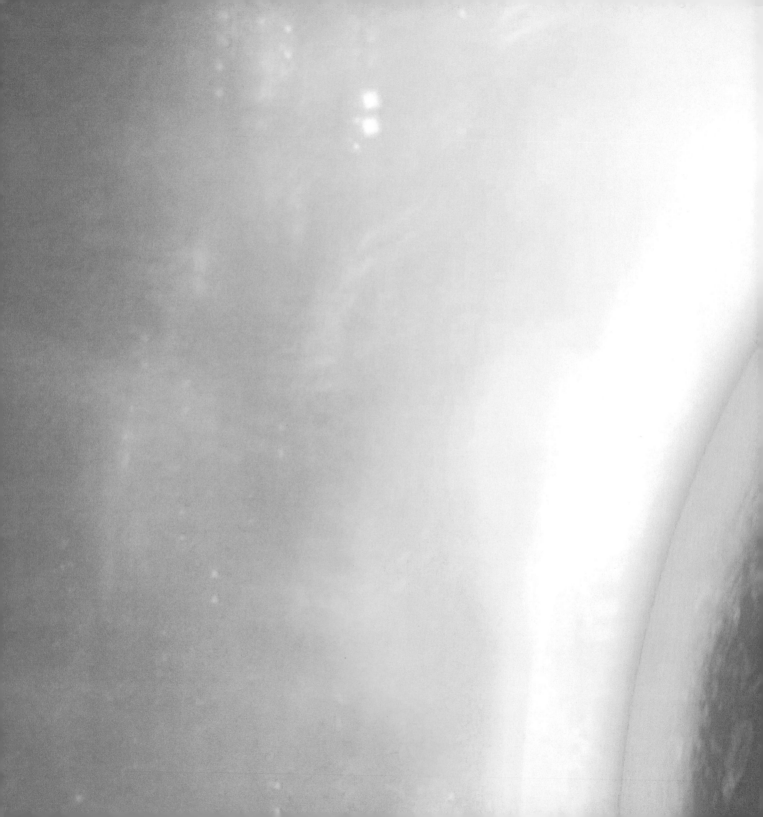

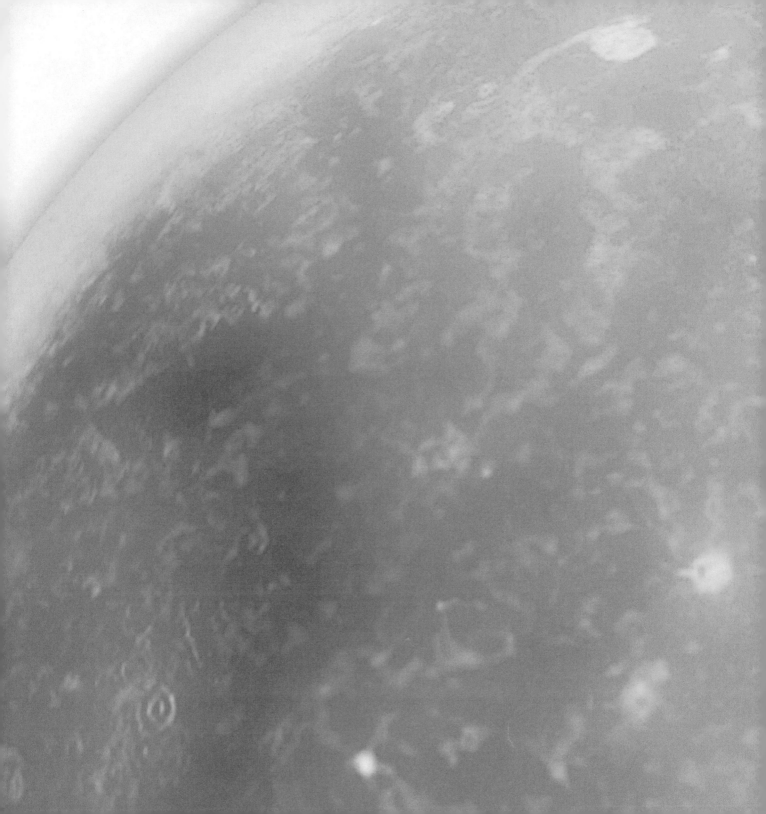

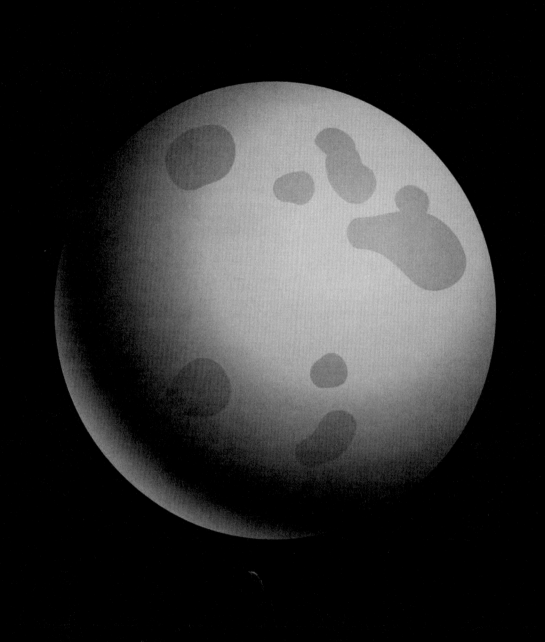

Did you know?

Mercury is the smallest and closest to the Sun of the eight planets in the Solar System, with an orbital period of about 88 Earth days. Seen from Earth, it appears to move around its orbit in about 116 days, which is much faster than any other planet. It has no known natural satellites. The planet is named after the Roman deity Mercury, the messenger to the gods.

Did you know?

Venus is the second planet from the Sun, orbiting it every 224.7 Earth days. It has no natural satellite. It is named after the Roman goddess of love and beauty. After the Moon, it is the brightest natural object in the night sky, reaching an apparent magnitude of 4.6, bright enough to cast shadows.

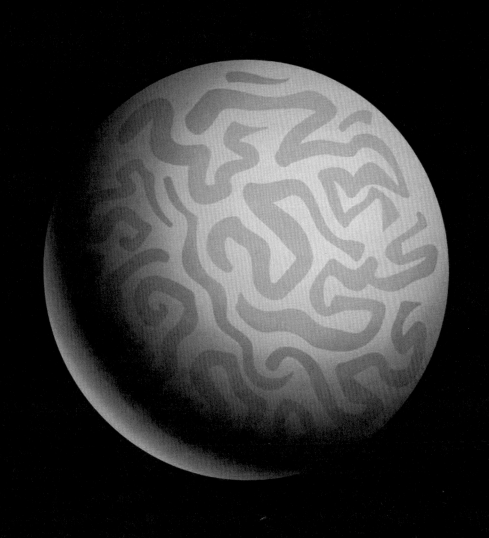

Did you know?

The Earth, also known as the world, Terra, or Gaia, is the third planet from the Sun, the densest planet in the Solar System, the largest of the Solar System's four terrestrial planets, and the only celestial body known to accommodate life. The Earth's biodiversity has evolved over hundreds of million years, expanding continually except when punctuated by mass extinctions. It is home to over eight million species.

Did you know?

Mars is the fourth planet from the Sun and the second smallest planet in the Solar System, after Mercury. Named after the Roman god of war, it is often described as the "Red Planet" because the iron oxide prevalent on its surface gives it a reddish appearance. Mars is a terrestrial planet with a thin atmosphere, having surface features reminiscent both of the impact craters of the Moon and the volcanoes, valleys, deserts, and polar ice caps of Earth.

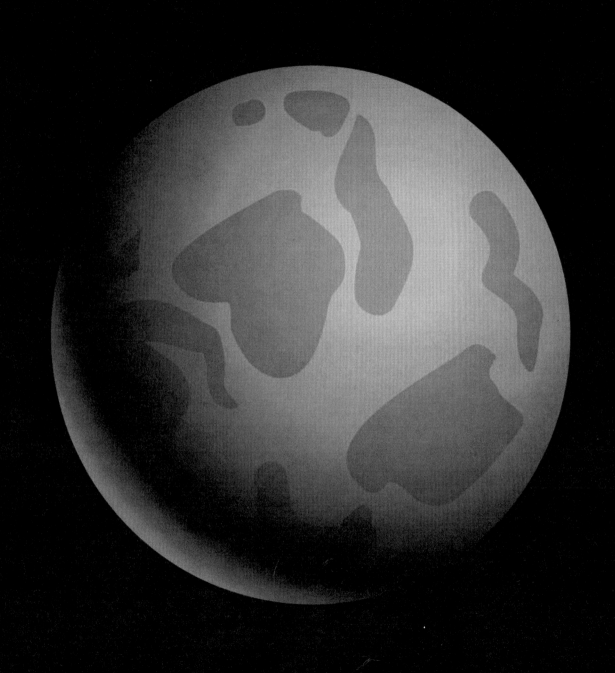

Did you know?

Jupiter is the fifth planet from the Sun and the largest planet in the Solar System. It is a gas giant with mass one-thousandth of that of the Sun but is two and a half times the mass of all the other planets in the Solar System combined. Jupiter is classified as a gas giant along with Saturn, Uranus and Neptune. Together, these four planets are sometimes referred to as the Jovian or outer planets. The planet was known by astronomers of ancient times. The Romans named the planet after the Roman god Jupiter.

Did you know?

Saturn is the sixth planet from the Sun and the second largest planet in the Solar System, after Jupiter. Named after the Roman god of agriculture, its astronomical symbol represents the god's sickle. Saturn is a gas giant with an average radius about nine times that of Earth. While only one-eighth the average density of Earth, with its larger volume Saturn is just over 95 times more massive.

Did you know?

Uranus is the seventh planet from the Sun. It has the third-largest planetary radius and fourth-largest planetary mass in the Solar System. Uranus is similar in composition to Neptune, and both are of different chemical composition to the larger gas giants Jupiter and Saturn. It is the only planet whose name is derived from a figure from Greek mythology rather than Roman mythology like the other planets, from the Latinized version of the Greek god of the sky, Ouranos.

Did you know?

Neptune is the eighth and farthest planet from the Sun in the Solar System. It is the fourth-largest planet by diameter and the third-largest by mass. Among the gaseous planets in the Solar System, Neptune is the most dense. Neptune is 17 times the mass of Earth and is slightly more massive than its near-twin Uranus, which is 15 times the mass of Earth but not as dense. Neptune orbits the Sun at an average distance of 30.1 astronomical units. Named after the Roman god of the sea.

Did you know?

A dwarf planet is an object the size of a planet (a planetary-mass object) but that is neither a planet nor a moon or other natural satellite. More explicitly, the International Astronomical Union (IAU) defines a dwarf planet as a celestial body in direct orbit of the Sun that is massive enough for its shape to be controlled by gravity, but that unlike a planet has not cleared its orbit of other objects.

Did you know?

It is estimated that there are hundreds to thousands of dwarf planets in the Solar System. The IAU currently recognizes five: Ceres, Pluto, Haumea, Makemake, and Eris. Pluto hit the headlines in 2006 when it was demoted from a planet to a dwarf planet by the International Astronomical Union. While most astronomers agreed with the new classifications, some disagreed and still refer to Pluto as the ninth planet.

Did you know?

Ceres is located in the asteroid belt between the orbits of Mars and Jupiter. It was discovered in 1801, well before Pluto and 45 years before Neptune. Ceres was considered a planet for around 50 years before being reclassified as an asteroid and once again in 2006 as a dwarf planet.

Did you know?

Eris was discovered in 2005 and was referred to as the tenth planet until it was reclassified in 2006. It is the largest of the dwarf planets.

Makemake was discovered in 2005 and the third largest dwarf planet behind Eris and Pluto.

Haumea was discovered in 2004 and named a dwarf planet in 2008.

Did you know?

The Moon (Latin: Luna) is Earth's only natural satellite. Although not the largest natural satellite in the Solar System, it is, among the satellites of major planets, the largest relative to the size of the object it orbits. It is the second-densest satellite among those whose densities are known (after Jupiter's satellite Io). The Moon orbits the Earth every 27.3 days. Ganymede is the largest moon in the Solar System. Neptune's largest moon, Triton, is similar in size to Earth's moon. Venus and Mercury have no moons.

Did you know?

As of 2009, there were 336 moons in the Solar System. 168 of these orbit planets, 6 orbit dwarf planets, while the rest orbit asteroids and other Solar System objects (many yet to be classified). Jupiter's four main moons are named the Galilean moons (after Galileo Galilei). Their names are Io, Europa, Ganymede and Callisto. Saturn's largest moon is named Titan, it is the only moon known to have a dense atmosphere. Mars has two moons named Phobos and Deimos, both were discovered in 1877. The largest moon of the dwarf planet Pluto is named Charon.

Made in the USA
San Bernardino, CA
26 November 2016